Bees

Science Kids
Life Cycles

Katie Gillespie

www.av2books.com

AV² provides enriched content that supplements and complements this book. Weigl's AV² books strive to create inspired learning and engage young minds in a total learning experience.

Your AV² Media Enhanced books come alive with...

 Audio Listen to sections of the book read aloud.

 Video Watch informative video clips.

 Embedded Weblinks Gain additional information for research.

 Try This! Complete activities and hands-on experiments.

 Key Words Study vocabulary, and complete a matching word activity.

 Quizzes Test your knowledge.

 Slide Show View images and captions, and prepare a presentation.

... and much, much more!

Go to www.av2books.com, and enter this book's unique code.

BOOK CODE

N287685

AV² by Weigl brings you media enhanced books that support active learning.

Published by AV² by Weigl
350 5th Avenue, 59th Floor New York, NY 10118
Website: www.av2books.com

Copyright ©2017 AV² by Weigl
All rights reserved. No part of this publication may be reproduced, stored in a retrieval system, or transmitted in any form or by any means, electronic, mechanical, photocopying, recording, or otherwise, without the prior written permission of the publisher.

Library of Congress Control Number: 2015958801

ISBN 978-1-4896-4499-2 (hardcover)
ISBN 978-1-4896-4500-5 (softcover)
ISBN 978-1-4896-4502-9 (multi-user eBook)

Printed in the United States of America in Brainerd, Minnesota
1 2 3 4 5 6 7 8 9 0 19 18 17 16 15

122015
041215

Project Coordinator: Jared Siemens
Art Director: Terry Paulhus

The publisher acknowledges Corbis Images, Minden Pictures, Alamy, and Getty Images as the primary image suppliers for this title.

Science Kids Life Cycles
Bees

CONTENTS

- 2 AV² Book Code
- 4 Insects
- 6 Life Cycle
- 8 Birth
- 10 Larvae
- 12 Pupae
- 14 The Change
- 16 Adult
- 18 Laying Eggs
- 20 Larvae to Bees
- 22 Life Cycles Quiz
- 24 Key Words

Bees are insects. Insects are small animals with six legs. Insects have skeletons on the outside of their bodies. These bodies have three parts.

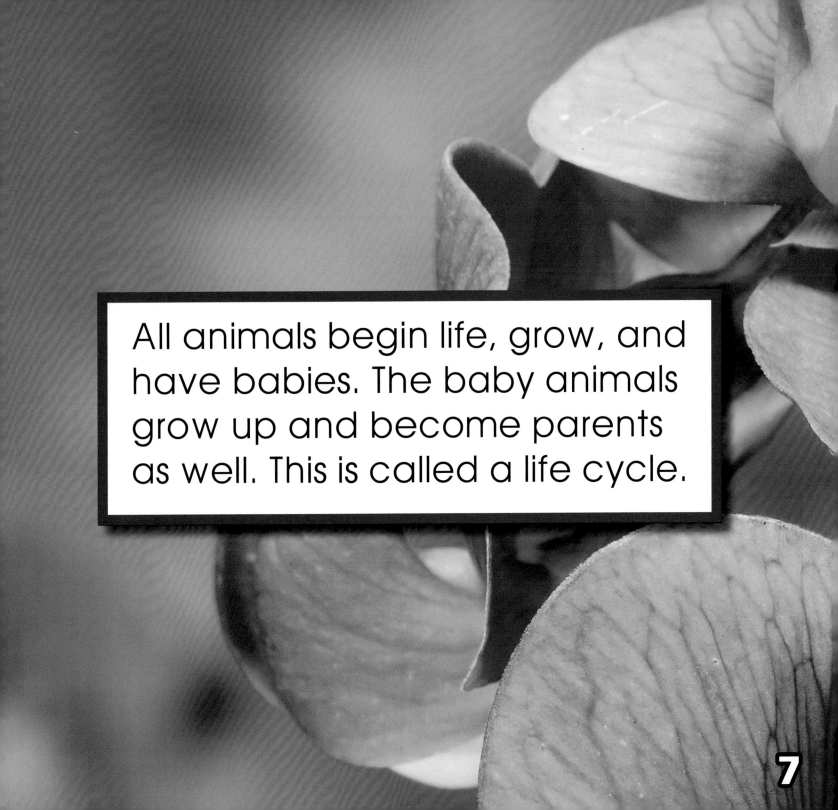

All animals begin life, grow, and have babies. The baby animals grow up and become parents as well. This is called a life cycle.

Bees are born when they hatch from eggs. These eggs look like tiny grains of rice.

Each egg is laid in its own place inside a beehive.

9

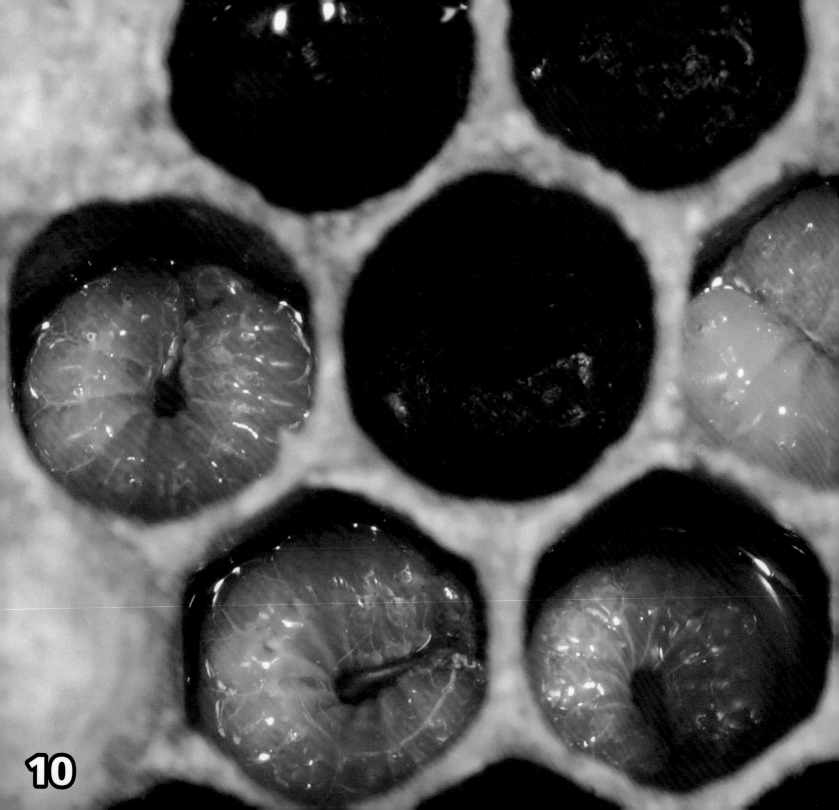

Baby bees are called larvae. They look like white worms. Larvae eat and grow very fast. Their old skin falls off their bodies as they grow larger.

Some bee larvae spin a cocoon when they grow big enough. This cocoon is made of silk. Bees are called pupae in this stage of the life cycle.

A pupa stays inside its cocoon for about 7 to 14 days.

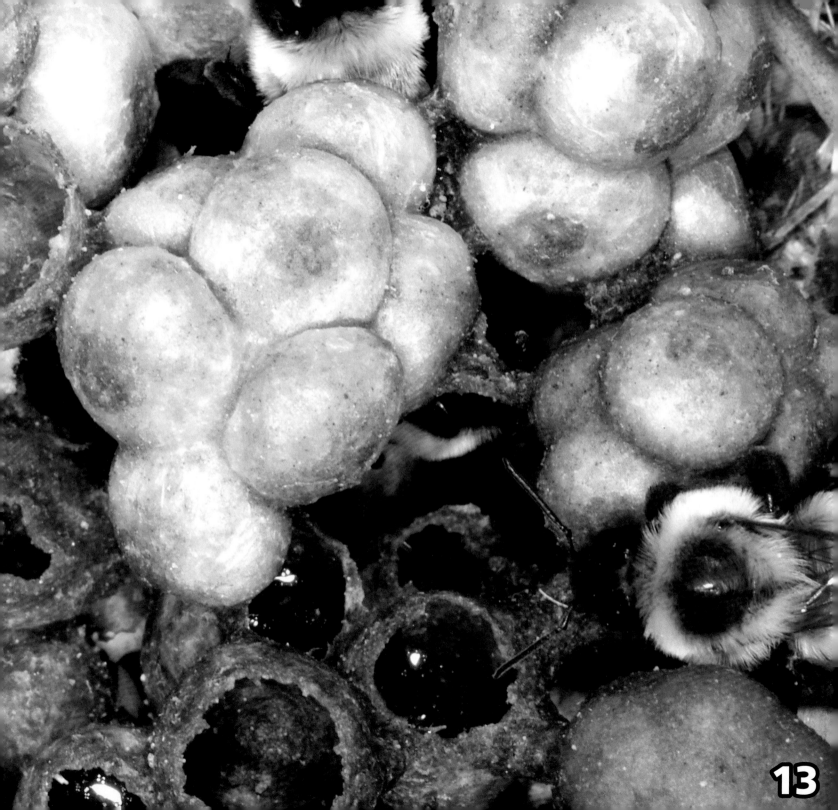

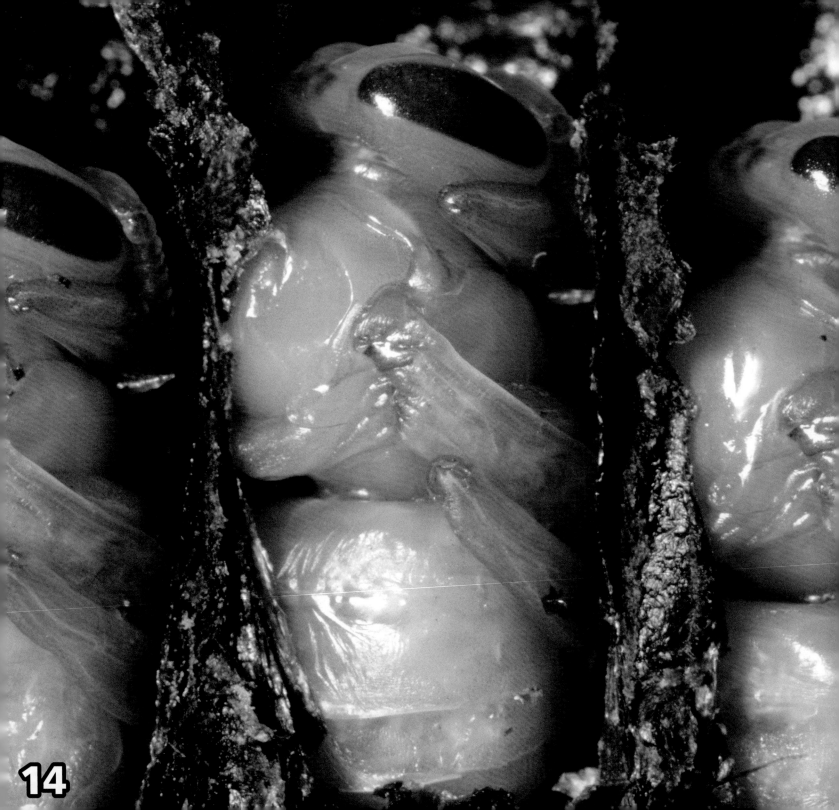

A bee's body changes shape in the pupa stage. Bees grow legs and eyes as they change into their adult form.

A bee is fully grown when it comes out of its cocoon. It has antennae and four wings. This is the adult stage of the life cycle.

A bee's life cycle takes about two to five weeks.

It is the queen bee's job to lay the eggs in a beehive. Some queens can lay up to 2,000 eggs each day.

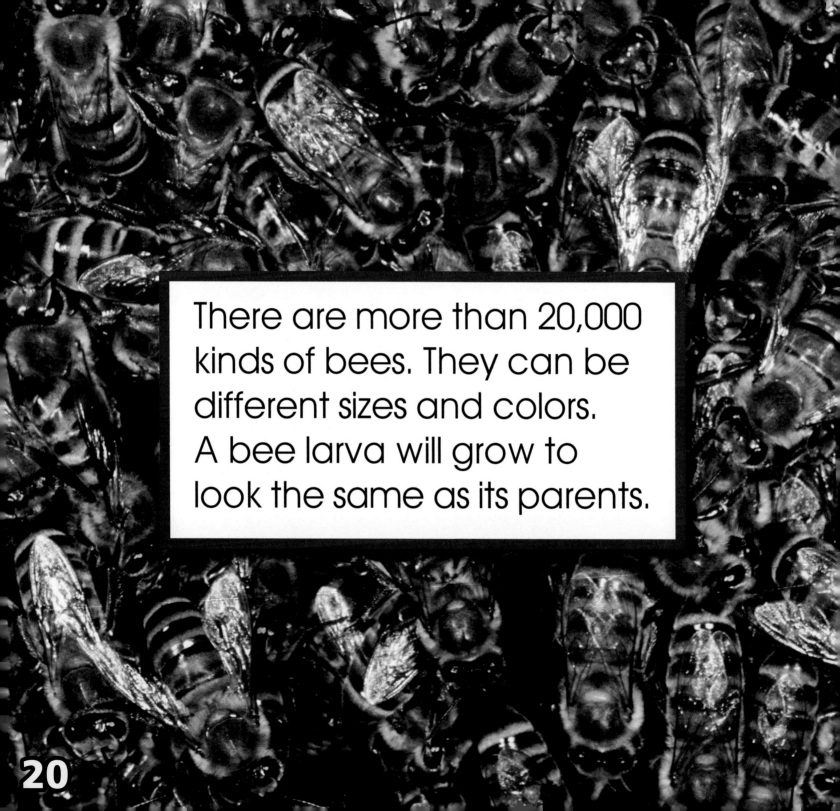

There are more than 20,000 kinds of bees. They can be different sizes and colors. A bee larva will grow to look the same as its parents.

Life Cycles Quiz

Test your knowledge of a bee's life cycle by taking this quiz. Look at these pictures. Which stage of the life cycle do you see in each picture?

Egg   Larva
Pupa   Adult

23

KEY WORDS

Research has shown that as much as 65 percent of all written material published in English is made up of 300 words. These 300 words cannot be taught using pictures or learned by sounding them out. They must be recognized by sight. This book contains 64 common sight words to help young readers improve their reading fluency and comprehension. This book also teaches young readers several important content words, such as proper nouns. These words are paired with pictures to aid in learning and improve understanding.

Page	Sight Words First Appearance
5	animals, are, have, of, on, parts, small, the, their, these, three, with
7	a, all, and, as, grow, is, life, this, up, well
8	each, from, in, its, like, look, own, place, they, when
11	eat, off, old, very, white
12	about, big, days, enough, for, made, some, to
15	changes, eyes, into
17	comes, four, has, it, out, takes, two
19	can
20	be, different, kinds, more, same, than, there, will

Page	Content Words First Appearance
5	bees, bodies, insects, legs, skeletons
7	babies, life cycle, parents
8	beehive, eggs, grains, rice
11	larvae, skin, worms
12	cocoon, pupae, silk
15	adult, shape
17	antennae, weeks, wings
19	job, queen bee
20	colors, sizes

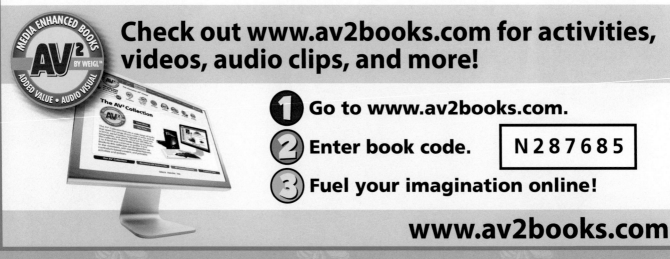

Check out www.av2books.com for activities, videos, audio clips, and more!

1. Go to www.av2books.com.
2. Enter book code. N287685
3. Fuel your imagination online!

www.av2books.com